NISTIR 7532

X-ray Microtomography Studies of Air-Void Instability and Growth during Drying of Tile Adhesive Mortars

Dale P. Bentz
Materials and Construction Research Division
Building and Fire Research Laboratory
National Institute of Standards and Technology
Gaithersburg, MD 20899-8615

Claus-Jochen Haecker
SE Tylose GmbH & Co. KG (Shin-Etsu Chemical)
Rheingaustrasse 190 - 196, Building H346
65203 Wiesbaden, GERMANY

November 2008

U.S. Department of Commerce
Carlos M. Gutierrez, Secretary

National Institute of Standards and Technology
Patrick D. Gallagher, Deputy Director

Abstract

A previous X-ray absorption study of tile adhesive mortars [1] identified a new phenomenon during drying, namely the movement of small cement particles from the interior of the specimen to its surface. In that study, an analysis based on Stokes' equation indicated that particle movement is consistent with the very high viscosity of these tile adhesive mortars. In addition to this particle movement, the concurrent formation/growth of large scale flaws (voids) within the specimens was observed. The goal of the present study was to observe this air void instability and growth in real time using three-dimensional X-ray microtomography. Mitigation strategies for avoiding the formation of these flaws were also investigated. Successful mitigation strategies included the utilization of a coarser cement (larger particles that do not move during drying), the addition of an accelerator (to achieve set before flaw formation can occur), and the addition of a co-thickener (to form a polymer network to prevent particle movement). Conversely, the addition of an air entrainer was not successful as a mitigation strategy.

Table of Contents

Abstract iii
List of Figures vi
List of Tables viii
1. Introduction 1
2. Materials and Experimental Methods 2
3. Results 6
4. Image Gallery from X-ray Microtomography Measurements 9
5. Conclusions 32
6. Acknowledgements 32
References 33

List of Figures

Figure 1. Measured particle size distributions for the two cement powders used in this study. The x-axis indicates an equivalent spherical particle diameter and the y-axis the mass % at each given diameter. The shown results are the average of six individual measurements and the error bars (one standard deviation) would fall within the size of the thicker line. 2

Figure 2. Measured mass loss during drying for the mortars prepared for this study. 4

Figure 3. Isothermal calorimetry results for the nine mortar mixtures prepared for this study. Top figure shows curves for mixtures with heat flow similar to the control mortar and those with acceleration. Bottom figure shows curves for mixtures with heat flow similar to the control and those with retardation.. 5

Figure 4. Diagram showing coronal, sagittal, and transaxial views for the cylindrical mortar specimens. 9

Figure 5. Coronal views of a cross-section of mortar #1 after various drying times, indicating instability and growth of large void near the top surface. Tile adhesive mortar specimens are approximately 11 mm in thickness. 10

Figure 6. Sagittal views of a cross-section of mortar #1 after various drying times, indicating instability and growth of large void near the top surface.. 11

Figure 7. Transaxial views of a cross-section of mortar #1 after various drying times. Transaxial images are 32.5 mm x 32.5 mm. 12

Figure 8. Coronal, transaxial, and sagittal images of mortar #2 after being exposed to drying for 30 min. 14

Figure 9. Coronal, transaxial, and sagittal images of mortar #2 after being exposed to drying for 1660 min. 15

Figure 10. Coronal, transaxial, and sagittal images of mortar #3 after being exposed to drying for 30 min. 16

Figure 11. Coronal, transaxial, and sagittal images of mortar #3 after being exposed to drying for 2475 min. 17

Figure 12. Coronal views of a cross-section of mortar #4 after various drying times, indicating instability and growth of large void near the right side. 18

Figure 13. Sagittal views of a cross-section of mortar #4 after various drying times. 19

Figure 14. Transaxial views of a cross-section of mortar #4 after various drying times.............. 20

Figure 15. Coronal, transaxial, and sagittal images of mortar #5 after being exposed to drying for 30 min. 22

Figure 16. Coronal, transaxial, and sagittal images of mortar #5 after being exposed to drying for 950 min. 23

Figure 17. Coronal, transaxial, and sagittal images of mortar #6 after being exposed to drying for 30 min. 24

Figure 18. Coronal, transaxial, and sagittal images of mortar #6 after being exposed to drying for 270 min. 25

Figure 19. Coronal, transaxial, and sagittal images of mortar #7 after being exposed to drying for 30 min. 26

Figure 20. Coronal, transaxial, and sagittal images of mortar #7 after being exposed to drying for 1040 min. 27

Figure 21. Coronal, transaxial, and sagittal images of mortar #8 after being exposed to drying for 30 min. 28

Figure 22. Coronal, transaxial, and sagittal images of mortar #8 after being exposed to drying for 260 min. 29

Figure 23. Coronal, transaxial, and sagittal images of mortar #9 after being exposed to drying for 30 min. 30

Figure 24. Coronal, transaxial, and sagittal images of mortar #9 after being exposed to drying for 1370 min. 31

List of Tables

Table 1. Mortar mixture proportions and measured air contents and open times............................ 3

1. Introduction

Specialty thin layer mortars are often formulated for use as tile adhesives or renderings for external and internal walls. Two common additives to such mortars are cellulose ethers (CE) and redispersible polymer powders (RPP) [2, 3]. CE retains water in the mortar for a sufficient time to allow proper cement hydration and is also employed to control the workability and increase the volumetric yield of the fresh mortars, first by acting as a thickener and second by entraining a considerable amount of air voids (about 25 % by volume). RPP are often added to improve mechanical properties of the hardened mortar. A key property of these mortars is their open time, defined as the time during which tiles may be applied to the mortar surface and achieve adequate adhesion (i.e., the time during which the mortar surface remains "tacky" to the touch) [3]. Ultimately, this tackiness is lost due to the formation of a film at the top surface of the mortars. The film may be composed of both polymeric materials and (carbonated) hydration products. Jenni et al. [2] used a variety of analytical techniques to study the polymer-mortar interactions in these materials. They observed, for example, that both the CE and the RPP are dissolved/redispersed in the pore solution and can form isolated films at interfaces during drying. They also identified a fractionation process in which polymer components (and calcium hydroxide) were enriched at the top drying surface of the mortars. Recently [1], X-ray absorption measurements have indicated that in addition to the polymer components and calcium hydroxide "concentrating" at the top surface during drying, small cement particles do likewise, due to the extremely high viscosity of the pore solution. Due to the additives present in these tile adhesive mortars, the pore solution viscosity becomes high enough to carry smaller particles along with the fluid during fluid movement from the interior of the specimen to the top (exposed) surface during drying.

The X-ray absorption measurements also indicated that particle movement during drying could be accompanied by the formation/growth of large scale (several mm in size) voids within the mortar microstructure. The formation of these large void regions could be partially responsible for the unusually low late age tensile strength sometimes exhibited by these materials, especially if the voids are formed at or near the interface between mortar and substrate. Voids of several mm in size could surely result in substantial strength reductions. Within the thin-set mortar industry, such voids have also been observed to produce anomalous needle penetration (setting) readings. In terms of open time, the densification of the top surface by the smaller particles in the system likely contributes to a reduction in the open time by further decreasing the "tackiness" of the top surface. Furthermore, their movement results in a concentration of the most reactive (smallest) cement particles at the top surface of the mortar where their more rapid hydration could further contribute to a reduction in open time both by reducing free water content and by stiffening the paste. Thus, from a practical standpoint, it would seem to be beneficial to formulate a mortar where this particle movement is minimized and air void instability and growth does not occur. In the present study, three-dimensional X-ray microtomography is employed to view in situ air void instability and growth in these tile adhesive mortars. A variety of mitigation strategies are evaluated on the basis of their ability to create a stable microstructure where air void instability is avoided.

2. Materials and Experimental Methods

Mortars were prepared in a 1.5 L plastic container using a single blade rotary mixer. All dry ingredients (cement, sand, CE, and other additives) were first homogenized in a sealed plastic bag. The dry ingredients were then placed in the mixing bowl and the appropriate mass of water added. Mixing proceeded for 1 min, followed by a rest period of 3 min, then 30 s of final mixing. The mixture proportions and air contents for the nine mortars are provided in Table 1. The mortars were proportioned on the basis of either 300 g or 150 g of dry ingredients (solids).

Two different binder powders were utilized in the experiments. For eight of the mortars, a European CEM I 42.5 N cement was used. For one of the experiments, a coarse cement based on a previous sieving (air classification) of Cement and Concrete Reference Laboratory proficiency sample 135 [4] was employed. The particle size distributions of the starting binder powders were measured using a laser diffraction method (Figure 1).

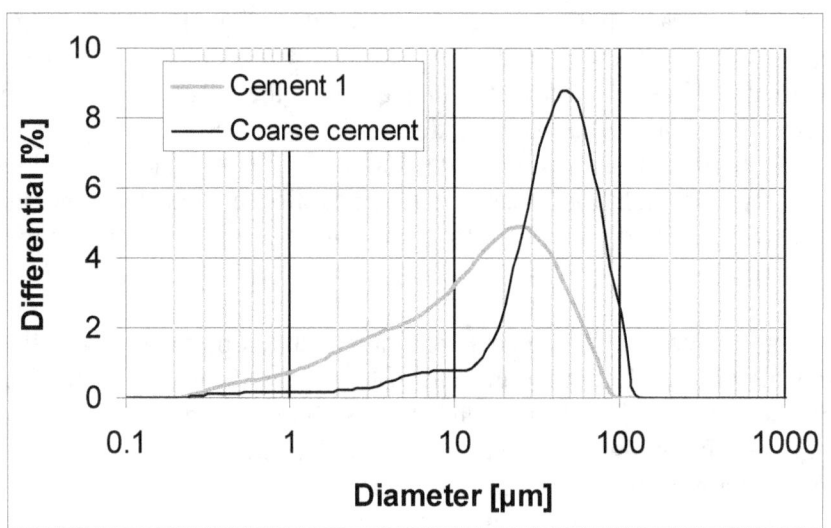

Figure 1. Measured particle size distributions for the two cement powders used in this study. The x-axis indicates an equivalent spherical particle diameter and the y-axis the mass % at each given diameter. The shown results are the average of six individual measurements and the error bars (one standard deviation) would fall within the size of the thicker line.

The prepared mortars were immediately placed into two round pre-weighed cylindrical plastic cups with a volume of 9130 mm^3 (inner diameter of 32 mm and height of about 11.4 mm) for the X-ray micro-tomography experiments and concurrent mass loss measurements. Initial masses of the filled cups, along with the volume of each, were used to estimate the air contents of the fresh mortars (Table 1). The average standard deviation in calculated air content between duplicate specimens was 0.9 %. Mass loss measurements were generally performed over the course of 1 d and were used to validate that the various prepared mortars were all drying at approximately the same rate (Figure 2). The temperature (nominally 23 °C) and relative humidity

Table 1. Mortar mixture proportions and measured air contents, open times, and peak heat flow times.

	Mortar 1	Mortar 2[*]	Mortar 3	Mortar 4	Mortar 5	Mortar 6	Mortar 7	Mortar 8	Mortar 9
Sand	191.7 g[+]	95.85 g	191.7 g	191.7 g	191.7 g	186.7 g	187.8 g	186.8 g	191.7 g
Cement	107.4 g	53.7 g[*]	107.4 g	107.4 g	107.4 g	104.6 g	105.3 g	104.6 g	107.4 g
Water	78 g	33 g	78 g	78 g	78 g	78 g	78 g	76.8 g	78 g
CE[&]	0.9 g	0.45 g	0.9 g	0.9 g	0.9 g	0.9 g	0.9 g	0.9 g	0.9 g
Co-thickener 1			0.18 g						
Co-thickener 2					0.18 g				
Air entrainer				0.24 g					
CaCl$_2$						7.8 g		7.7 g	
Polyvinyl alcohol (PVA)							6 g		
Starch ether									0.27 g
Imbentin[$]								1.6 g	
w/c	0.726	0.615	0.726	0.726	0.726	0.746	0.741	0.734	0.726
Air content	31.2 %	34.1 %	30.1 %	31.1 %	33.1 %	21.6 %	37.0 %	26.9 %	32.3 %
Open time	25 min	5 min[#]	10 min	25 min	25 min	80 min	15 min	35 min	25 min
Peak heat flow time	12 h	20.7 h	22.3 h	13 h	12.7 h	2.8 h	11.9 h	2.7 h	41.5 h
Stability	Unstable	Stable	Stable	Unstable	Stable	Stable	Stable	Stable	Unstable

[*] Coarse cement used in mortar mixture #2 only.
[+] The mass of each ingredient was measured to the nearest 0.01 g.
[&] Unmodified cellulose ether; methylhydroxyethylcellulose; high degree of etherification.
[$] Imbentin (Imbentin-AGS/35, 25 % in water) is an alcohol ethoxylate.
[#] Measured in a previous experiment [1], not as part of the current study.

(nominally 50 %) inside the X-ray microtomography chamber and in the laboratory where the drying specimens were stored were monitored using a USB-compatible data logger throughout the course of the experiment. The mass loss results in Figure 2 indicate that within experimental variability, the nine mortar mixtures exhibited similar mass loss curves during drying.

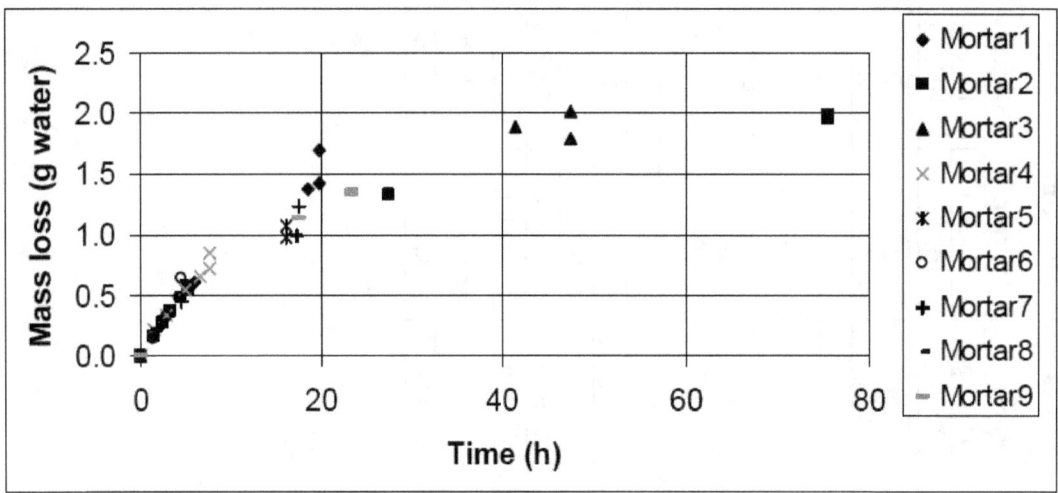

Figure 2. Measured mass loss during drying for the mortars prepared for this study.

Hydration rates were monitored using isothermal calorimetry on two replicate mortar specimens, each with a mass of 5.9 g. The heat flow curves for the two replicates generally overlapped one another so that only a single curve for each mixture is presented in the results in Figure 3. For each curve in Figure 3, the time of peak (maximum) heat flow is included in Table 1 to provide an indication of any acceleration or retardation produced by the additives relative to the "control" mortar (#1). Based on these values, the heat flow curves in Figure 3 clearly indicate the successful acceleration of the hydration by the addition of calcium chloride (mortars #6 and #8), the retardation of hydration by various additives (starch ether in mortar #9 and co-thickener in mortar #3), and the very low hydration rates exhibited by the coarse cement (mortar #2).

X-ray microtomography was conducted using a Skyscan 1172 system.[1] Operating parameters consisted of a voltage of 90 keV, a current of 112 μA, the utilization of aluminum and copper filters, and a resolution (voxel size) of 32 μm. These operating conditions were held constant for all of the three-dimensional image sets that were acquired during this study. Once the specimen for a given mortar was placed in the X-ray microtomography chamber, images were then acquired at multiple times up to 24 h without any movement of the specimen or opening/closing of the chamber door. A single complete three-dimensional scan took 10 min. Three-dimensional images were reconstructed from the raw scan data using the system software. Representative images obtained at the same (x,y,z) location but after different drying times were

---

[1] Certain commercial products are identified in this paper to specify the materials used and procedures employed. In no case does such identification imply endorsement by the National Institute of Standards and Technology, nor does it indicate that the products are necessarily the best available for the purpose.

then compared to determine the stability/instability of the air void system in each mortar during drying.

In separate experiments, the open times of the various mortars were measured according to European Standard EN 1347 (Determination of the wetting capability) [5]. Using this test method, a layer of a tile adhesive is combed onto a concrete slab using a notched trowel. Glass plates (100 mm × 100 mm × 6 mm) are embedded in the fresh mortar after 5 min, 10 min, 15 min, etc., from which the ability of the applied material to wet the glass plates is assessed. Only samples with a wetted area > 50 % pass the test. The open time results are included in Table 1.

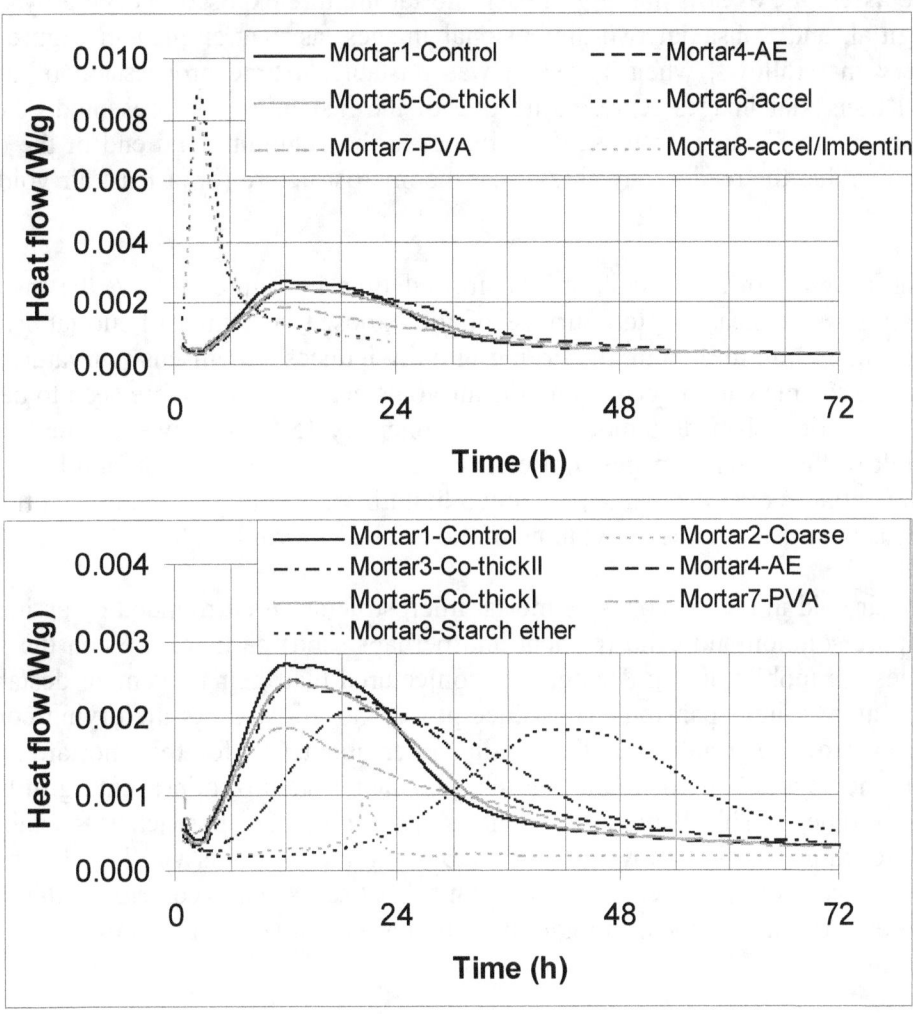

Figure 3. Isothermal calorimetry results for the nine mortar mixtures prepared for this study. Top figure shows curves for mixtures with heat flow similar to the control mortar and those with acceleration. Bottom figure shows curves for mixtures with heat flow similar to the control and those with retardation.

3. Results

The goals of the present study were twofold: 1) to observe air void instability and growth during the drying of tile adhesive mortars to verify that previously observed large scale flaws had not been present immediately after mixing and casting, and 2) to determine the effectiveness of various mitigation strategies in preventing this instability and growth. With this in mind, the control mortar (#1) was selected as a likely candidate in which to observe this instability, based on the results of the previous X-ray absorption experiments [1]. Once the complete three-dimensional data sets were obtained for a given experiment, data viewing software was employed to assess the overall stability of each mortar mixture exposed to drying by viewing the coronal, sagittal, and transaxial two-dimensional images, as diagrammed in Figure 4. In the image gallery that follows, when a system was unstable, a time progression of the coronal, sagittal, and transaxial images representing one of the instabilities is presented. For a stable system, representative images in these three planes at the beginning and end of the experiment are presented so that the reader may assess that the microstructure (or at least air void structure) is indeed stable.

As an example of an unstable mortar formulation, in Figures 5 to 7, the instability and growth of an air void near the top surface of the exposed specimen of mortar #1 is clearly observed. From 30 min to 90 min, the drying initially induces a meniscus (curvature) at the top surface as some sedimentation occurs, and the air voids near the surface are seen to deform from their initially spherical shape to a more ellipsoidal one. By 150 min, however, the large air void on the top left of the coronal images of the specimen has become unstable and begins to grow. This growth continues over the course of about 3 h, until set occurs, ultimately resulting in a flaw (void) that has more than double the volume of the original air void.

Each air void in the tile adhesive mortar microstructure is surrounded by an interface that consists of pore solution and solid (cement and perhaps sand) particles. When the smallest of these particles are mobile during drying, it is conjectured that their movement destabilizes this interface and allows the expansion/coalescence process to initiate. As the drying continues, so does the growth of the instability. Eventually, after about 6 h for this mortar, initial set is achieved and the physical microstructure of the specimen is stabilized, preventing further growth of these instabilities. This hypothesis suggests three possible approaches to stabilizing the system: 1) preventing particle movement, 2) stabilizing the air voids via the addition of an air entrainer, or 3) accelerating the cement hydration so that set is achieved prior to the initiation of the instabilities. All three of these mitigation strategies were investigated in the eight subsequent mortar mixtures.

In the 2nd experiment, the control cement was replaced by a much coarser cement that had been obtained by classifying an ordinary portland cement in a previous study [4]. The previous X-ray absorption study [1] had indicated that this was a viable approach to minimizing particle movement during drying. Indeed, in this study, no evidence of instabilities could be observed in the X-ray microtomography data sets. Figures 8 and 9 provide representative images at the beginning (30 min) and end of the experiment (1660 min); any differences between the two sets of images are quite minimal. Unfortunately, previous results [1] have indicated a very short open time of only 5 min for this system (there was not enough cement remaining to measure the open time in this study) and as shown in Figure 3, hydration achieved during the first 72 h is minimal in comparison to that achieved by the control cement. The minimal hydration indicated for this mortar is due to both the large size of the cement particles and the fact that this separated coarse cement had been stored in a closed container for over 5 years.

For the 3rd mortar mixture, a co-thickener was utilized along with the cellulose ether. As shown in Figures 10 and 11, this produced a stable microstructure. In viewing the transaxial images for these two figures, some changes are observed, but these are due to settlement of the specimen (as noted in the corresponding coronal and sagittal images). Viewing the isothermal calorimetry results in Figure 3, it is observed that this co-thickener caused a measurable retardation of the cement hydration reactions. This would likely correspond to an increase in the setting time for this particular mortar, which could possibly be an advantage in increasing the open time or a disadvantage in slowing down strength development. However, separate measurements of the open time for this mortar instead indicated a reduction from the 25 min measured on the control mortar #1 to only 10 min, perhaps due to enhanced film formation at the top surface due to the addition of the co-thickener.

For the 4th mortar mixture, an air entrainer was added in an attempt to (chemically) stabilize the air voids without eliminating the particle movement. As observed in the coronal, sagittal, and transaxial images in Figures 12, 13, and 14, respectively, this system exhibited the instability and growth of a large air void near the right edge (coronal view) of the specimen. For this study, stabilizing the air voids using an air entrainer was thus an unsuccessful mitigation strategy. This mortar exhibited the same open time as the control (Table 1).

For the 5th mortar mixture, a different co-thickener was employed, one that has minimal effects on the cement hydration reactions (see Figure 3). As observed in Figures 15 and 16, this also proved to be a successful mitigation strategy to avoid the formation and growth of instabilities, and provided the additional advantage of an open time equal to that of the control.

In the 6th mortar mixture, an accelerator (calcium chloride) was added at a fairly high dosage (7.5 % by mass of cement) to decrease the setting time and try to mitigate instabilities by achieving set prior to their formation. As indicated in Figure 3, the calcium chloride greatly accelerated the hydration reactions and the peak hydration rate was indeed achieved prior to 3 h. Figures 17 and 18 indicate that this strategy was also successful in avoiding instability formation and growth. However, it can be observed in Table 1 that the calcium chloride functioned somewhat as an air detrainer and reduced the air content of mortar #6 down to about 22 %, about 10 % less than that of the control mortar. The potential to offset this air content reduction by the incorporation of an additional air entrainer was investigated in the 8th mortar mixture to follow. Even though the calcium chloride produced a considerable acceleration of the hydration reactions, the measured open time for mortar #6 was 80 min (Table 1), much greater than that of the control mortar. The reason for this increased open time is unknown, but is perhaps due to favorable interactions between the Ca^{++} and Cl^- ions and the cellulose ether.

In the 7th mortar mixture, a polyvinyl alcohol was added to the control mortar in hopes of forming a polymeric network that would prevent the movement of the smaller cement particles during drying. This mortar exhibited a minor decrease in heat release (Figure 3), but was stable from a microstructural viewpoint (Figures 19 and 20), even though it contained some quite large (several mm in size) air voids initially. This mortar also exhibited the highest air content (greatest yield) of any of the mixtures investigated in this study (Table 1). The polyvinyl alcohol addition reduced the open time of the mortar from 25 min to only 15 min.

In the 8th mortar mixture, the calcium chloride accelerator and an air entrainer (Imbentin) were both added to the control mortar. This increased the air content from about 22 % to 27 %, still somewhat less than that of the control mortar. The heat release curves for mortars #6 and #8 (Figure 3) were quite similar and the stability of the mortar was verified, as shown in Figures 21 and 22. The measured open time, 35 min, exceeded that of the control mortar #1 by 10 min.

In the 9th mortar mixture, a starch ether was added to attempt to stabilize the microstructure by preventing particle movement during drying. As can be seen in Figure 3, this produced extensive (greater than 24 h) retardation of the cement hydration and as shown in Figures 23 and 24, resulted in a highly unstable microstructure in which void instability and growth (and even subsurface cracking) were clearly present. The measured open time of this mortar was equivalent to that of the control mortar #1.

In the unstable mortars (#1, #4, and #9) observed in this study, the instabilities, once initiated, were observed to generally grow in a downward direction. This could be influenced by the differential stresses/strains that are established through the thickness of the specimen due to water content, particle density, and degree of hydration all being a function of height within the specimen. The top of the specimen where the smaller particles concentrate may well have a higher local density [1] and a greater degree of cement hydration (since smaller particles have more reactive surface area for early-age hydration), both of which could contribute to it being better able to resist the stresses and strains causing the formation and growth of the instability than the material present at a lower depth within the specimen.

4. Image Gallery from X-ray Microtomography Measurements

In the images that follow, the views are labeled as coronal, sagittal, or transaxial, respectively, in agreement with the nomenclature advocated by the equipment manufacturer. As shown in Figure 4, the coronal view is defined as the plane perpendicular to the axis running from the X-ray source to the camera detector, with reference to the rotating table being at its initial position of 0 degrees. The sagittal view is at 90 ° relative to the coronal view, while the transaxial view is effectively looking directly down from the top into the specimen.

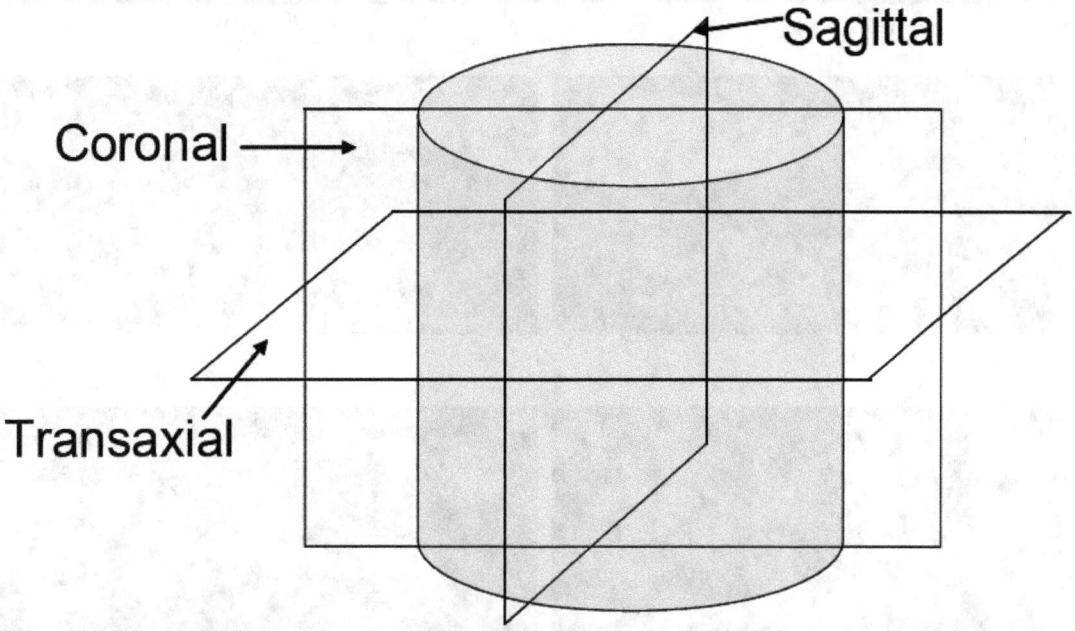

Figure 4. Diagram showing coronal, sagittal, and transaxial views for the cylindrical mortar specimens.

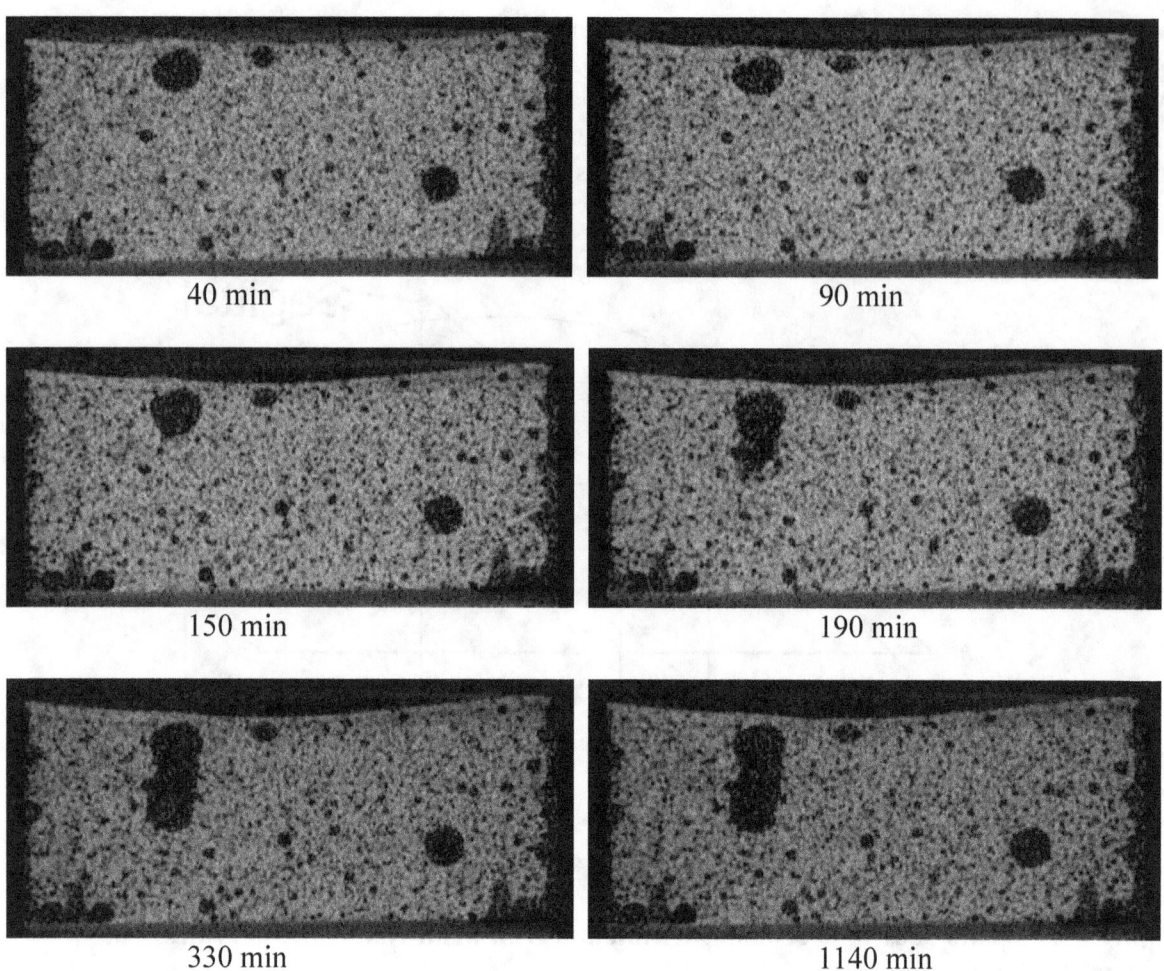

Figure 5. Coronal views of a cross-section of mortar #1 after various drying times, indicating instability and growth of large void near the top surface. Tile adhesive mortar specimens are approximately 11 mm in thickness.

Mortar #1 (Control) – Sagittal Views

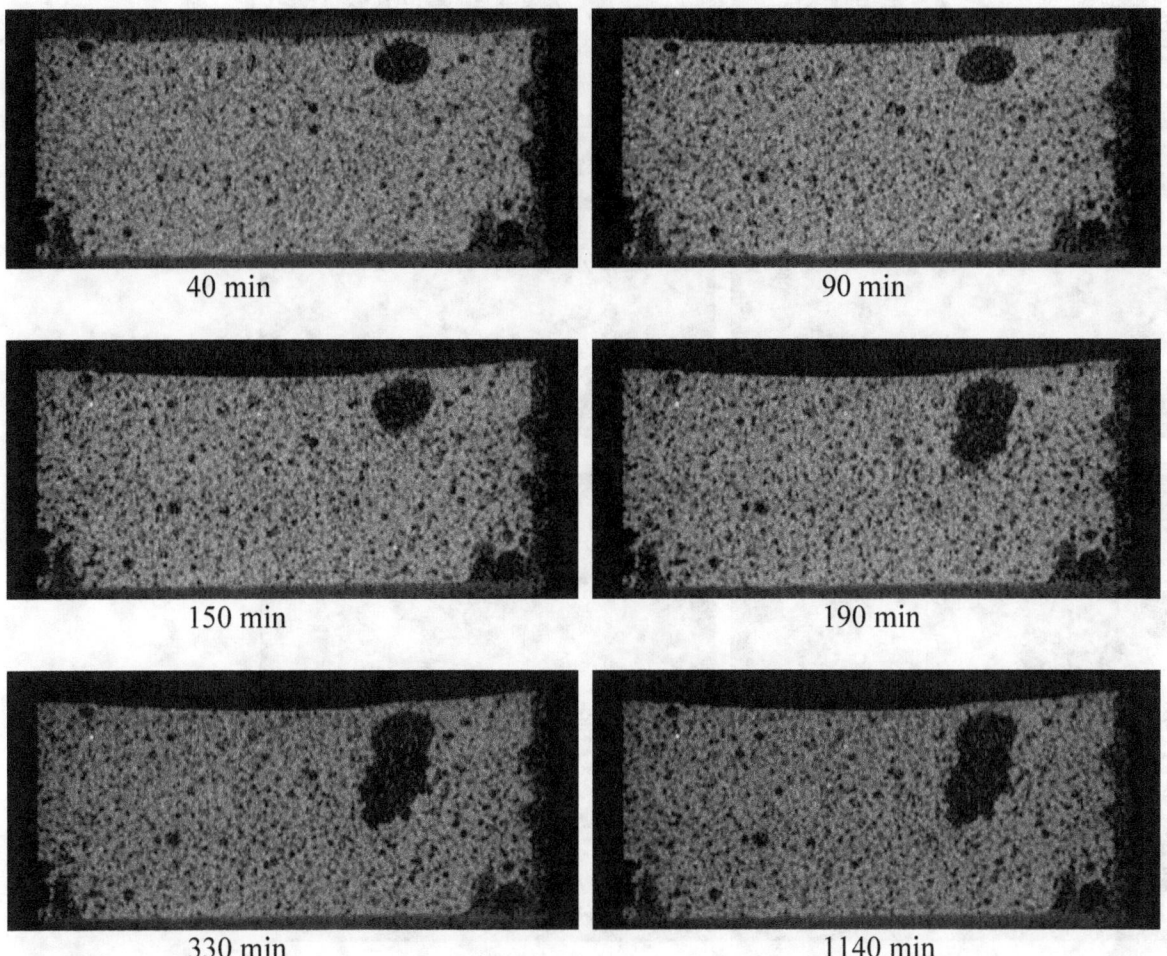

Figure 6. Sagittal views of a cross-section of mortar #1 after various drying times, indicating instability and growth of large void near the top surface.

Mortar #1 (Control) – Transaxial Views

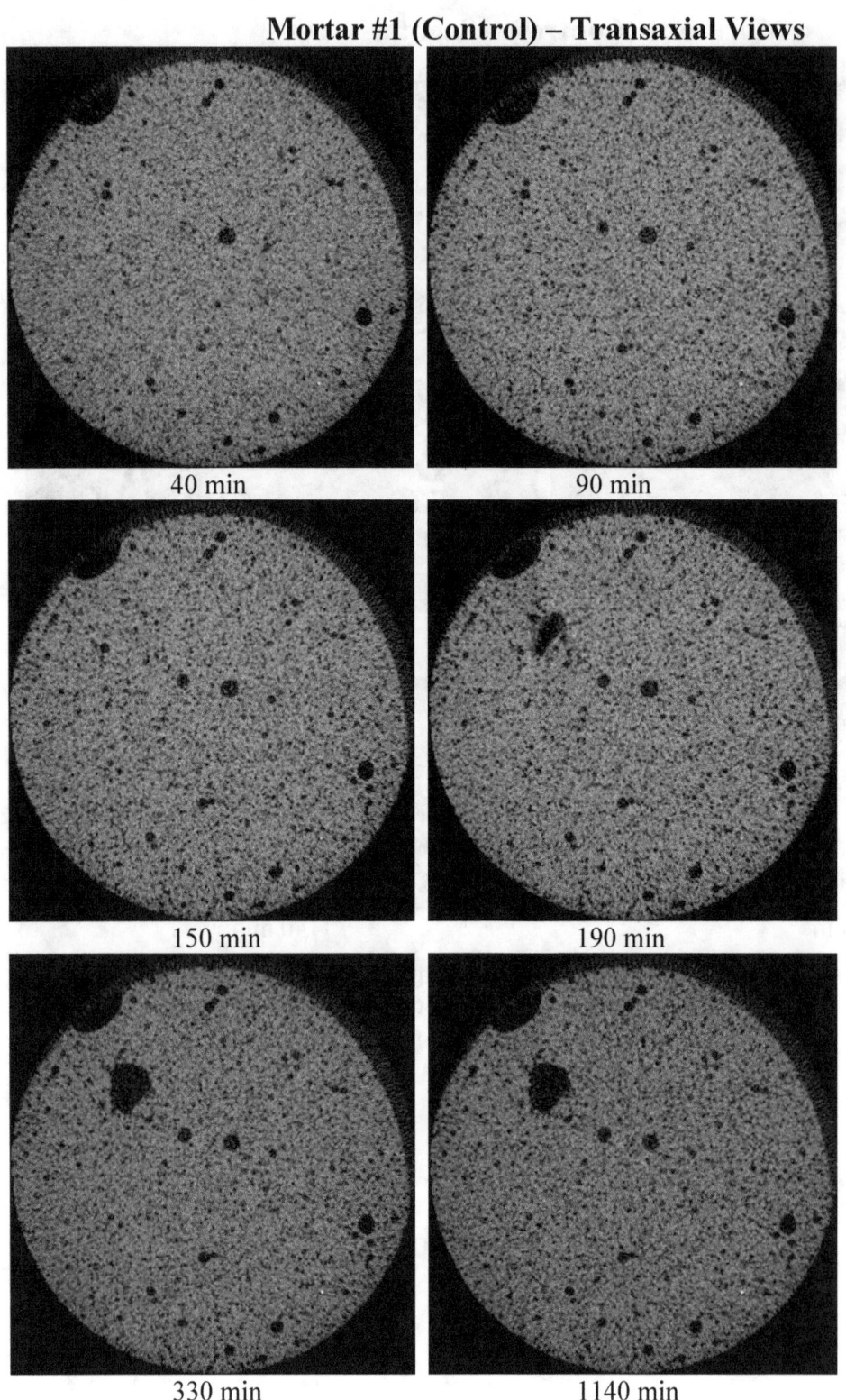

Figure 7. Transaxial views of a cross-section of mortar #1 after various drying times. Transaxial images are 32.5 mm x 32.5 mm.

Mortar #2 (Coarse cement) at 30 min

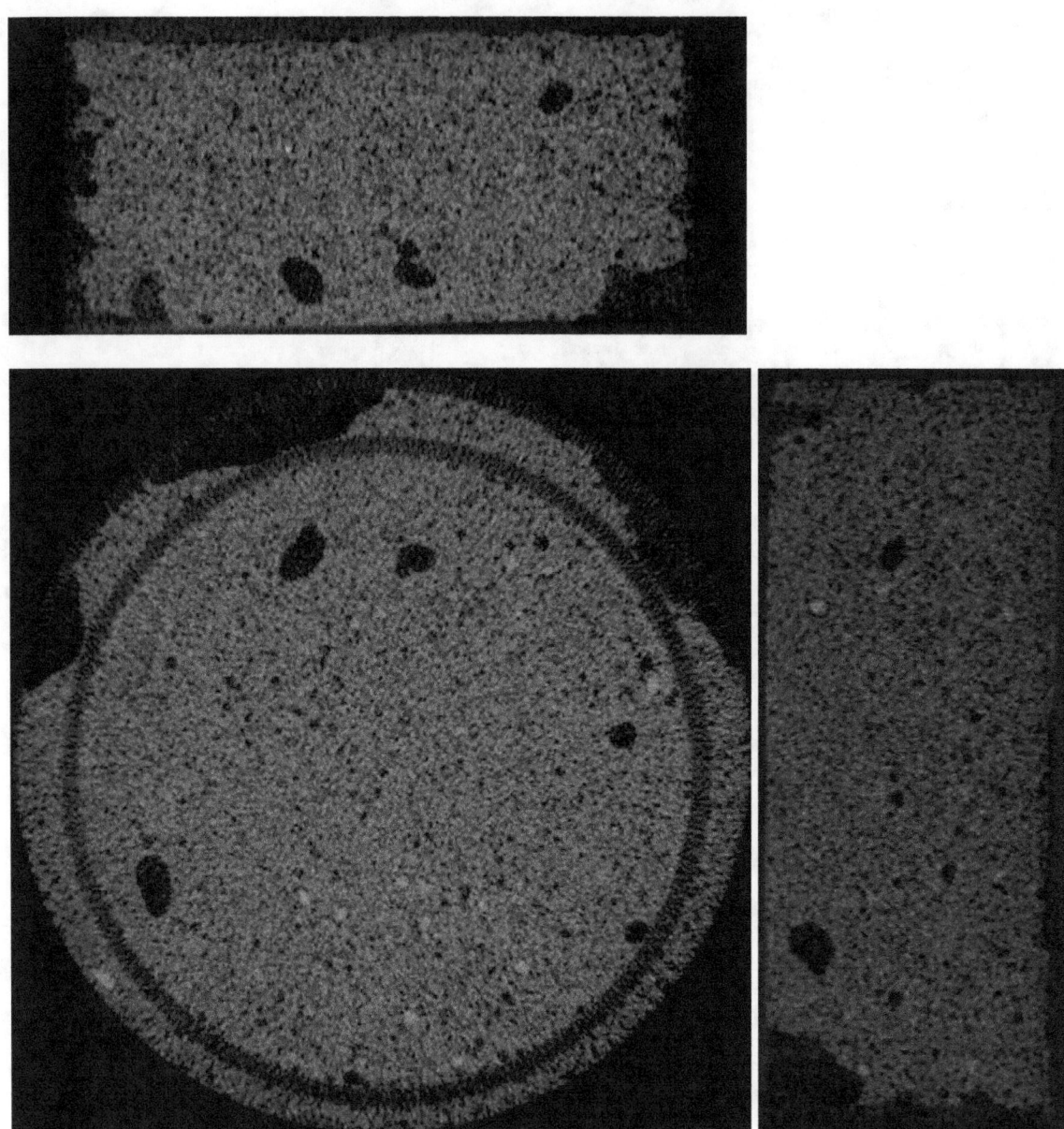

Figure 8. Coronal, transaxial, and sagittal images of mortar #2 after being exposed to drying for 30 min.

Mortar #2 (Coarse cement) at 1660 min

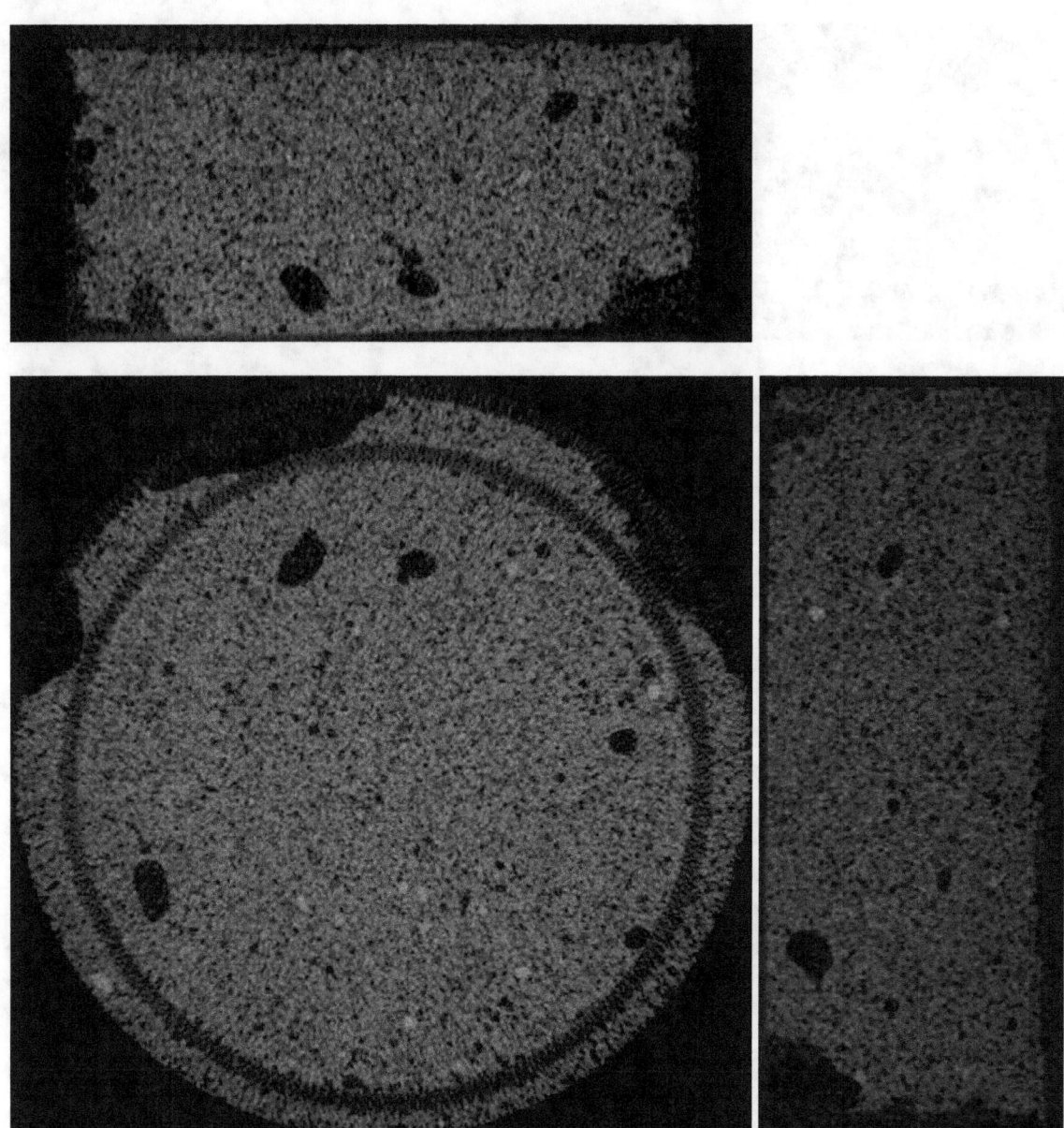

Figure 9. Coronal, transaxial, and sagittal images of mortar #2 after being exposed to drying for 1660 min.

Mortar #3 (Co-thickener II) at 30 min

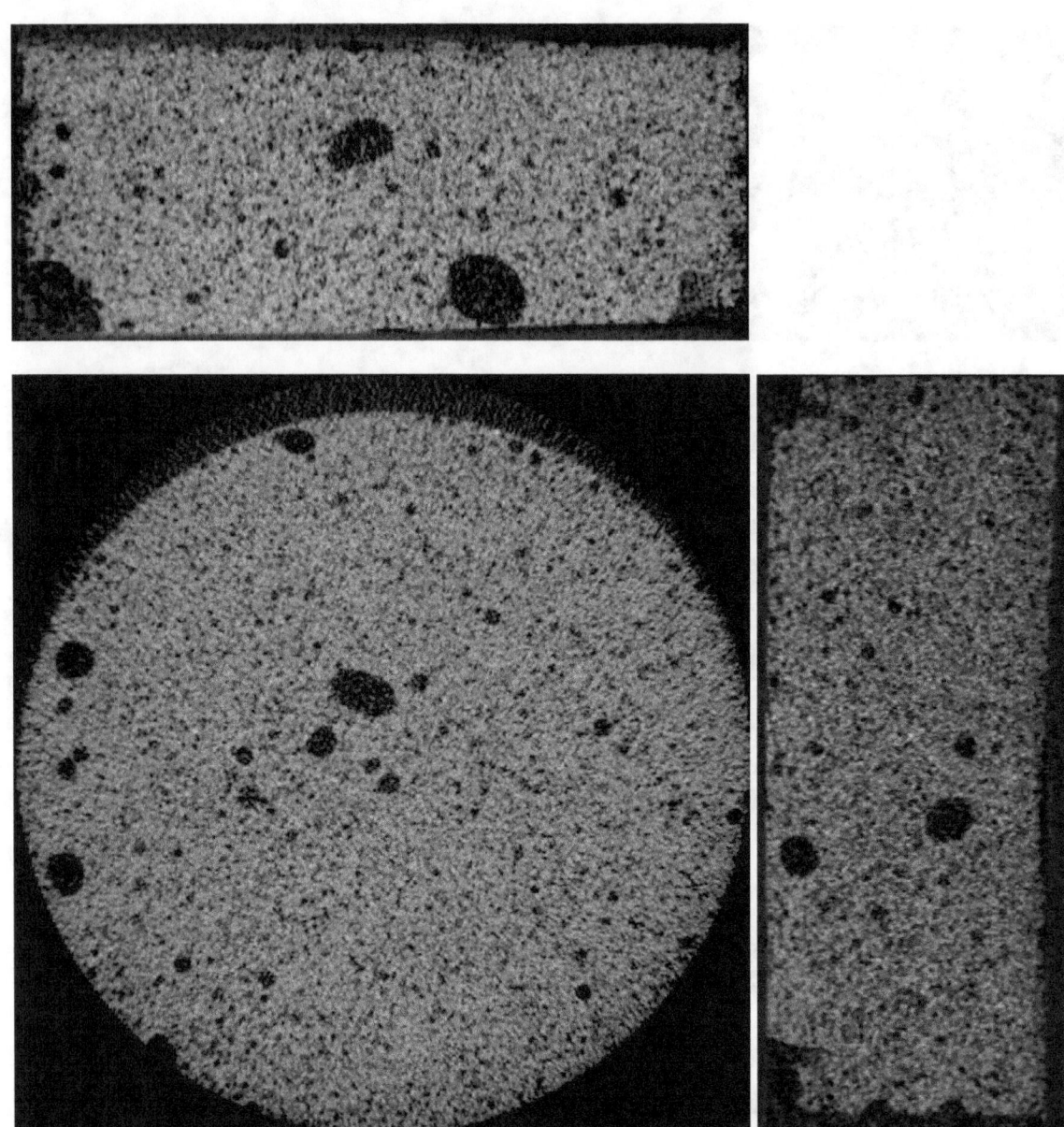

Figure 10. Coronal, transaxial, and sagittal images of mortar #3 after being exposed to drying for 30 min.

Mortar #3 (Co-thickener II) at 2475 min

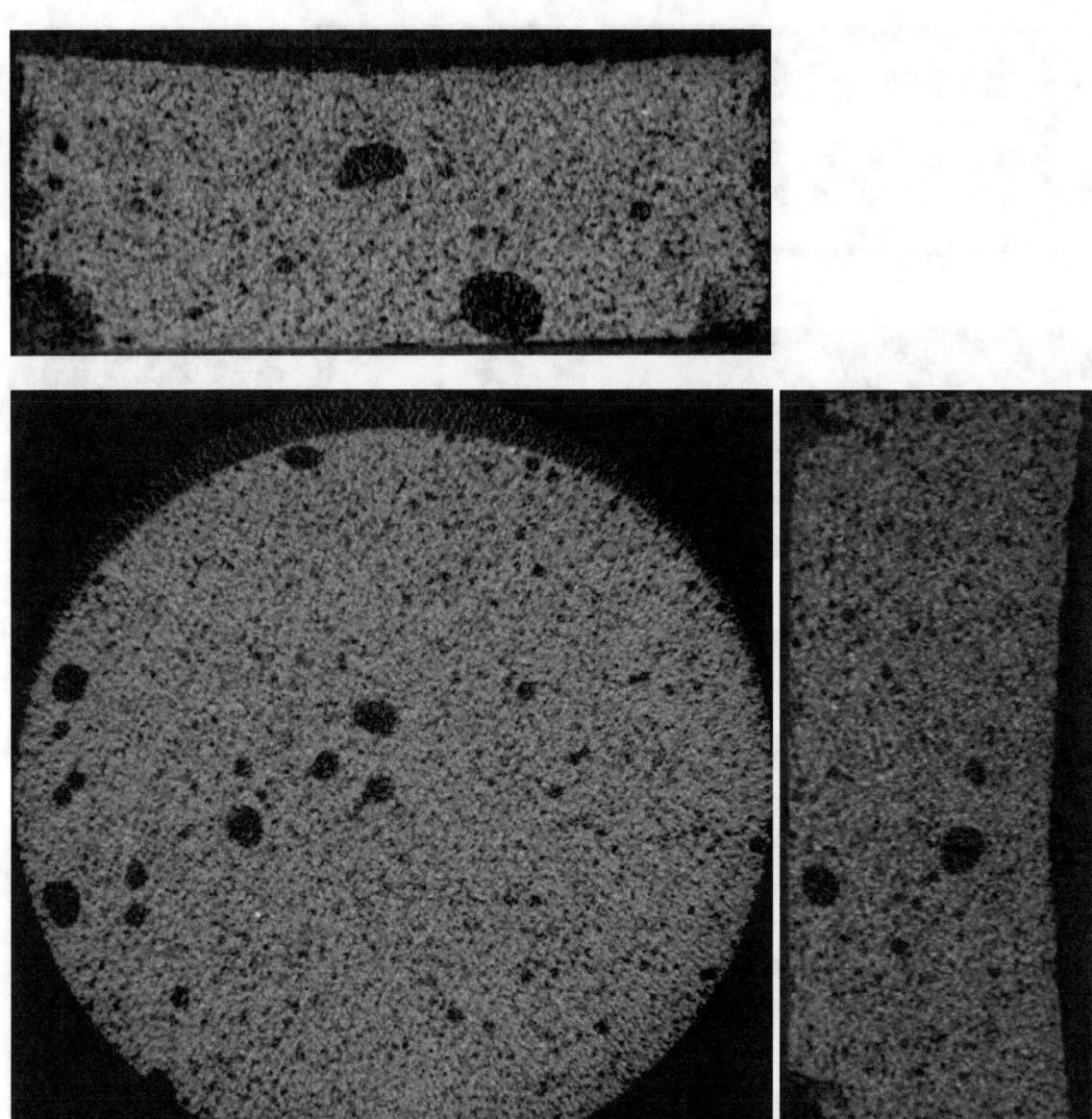

Figure 11. Coronal, transaxial, and sagittal images of mortar #3 after being exposed to drying for 2475 min.

Mortar #4 (Air entrainer) – Coronal Views

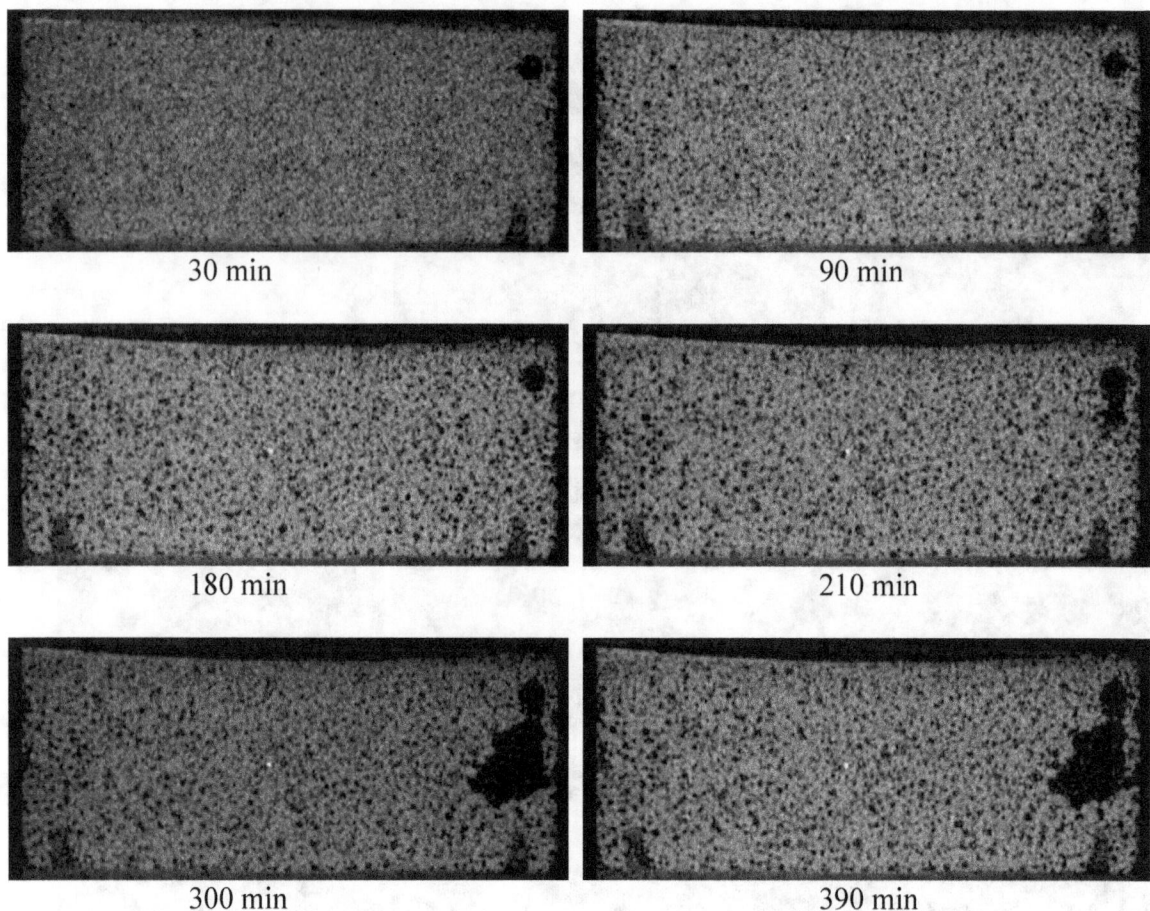

Figure 12. Coronal views of a cross-section of mortar #4 after various drying times, indicating instability and growth of large void near the right side.

Mortar #4 (Air entrainer) – Sagittal Views

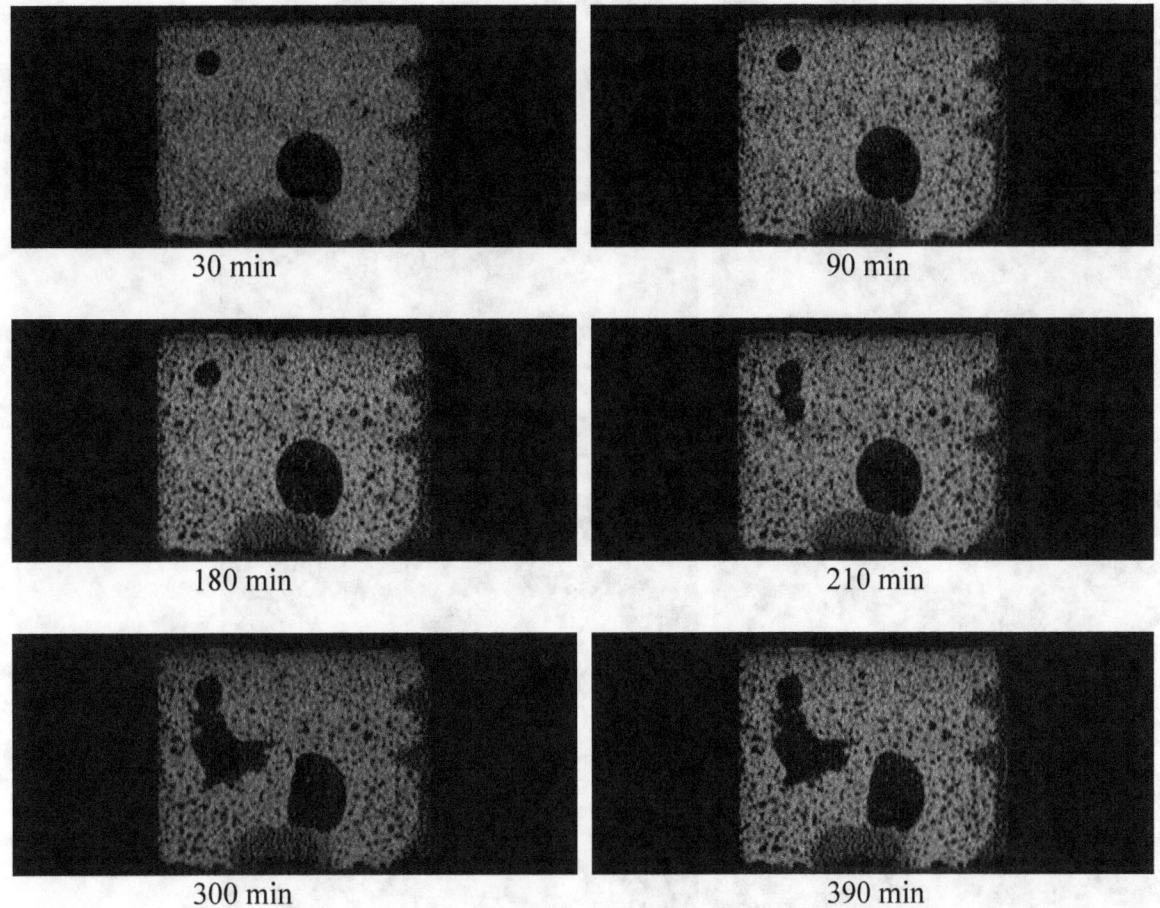

Figure 13. Sagittal views of a cross-section of mortar #4 after various drying times.

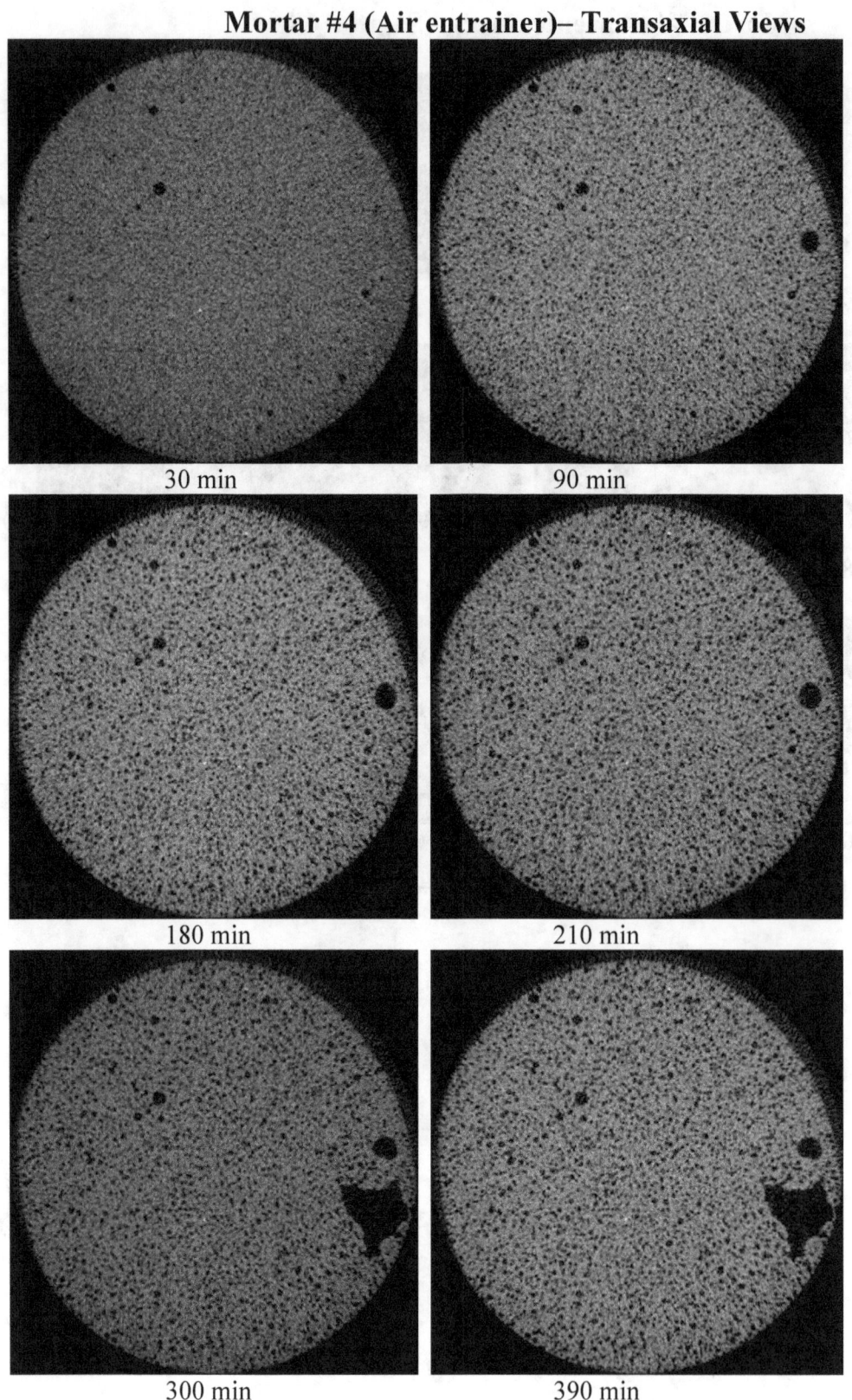

Figure 14. Transaxial views of a cross-section of mortar #4 after various drying times.

Mortar #5 (Co-thickener I) at 30 min

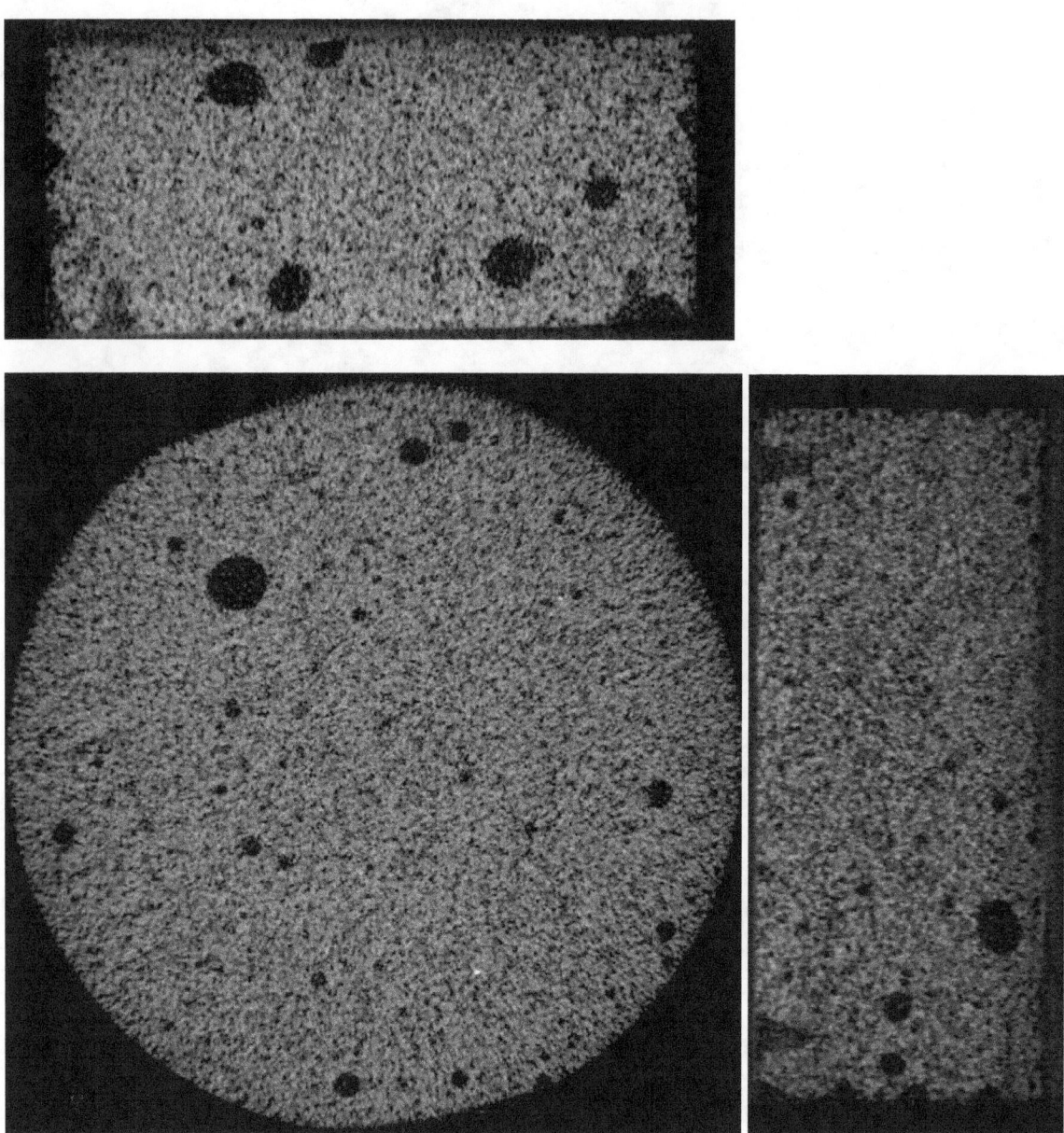

Figure 15. Coronal, transaxial, and sagittal images of mortar #5 after being exposed to drying for 30 min.

Mortar #5 (Co-thickener I) at 950 min

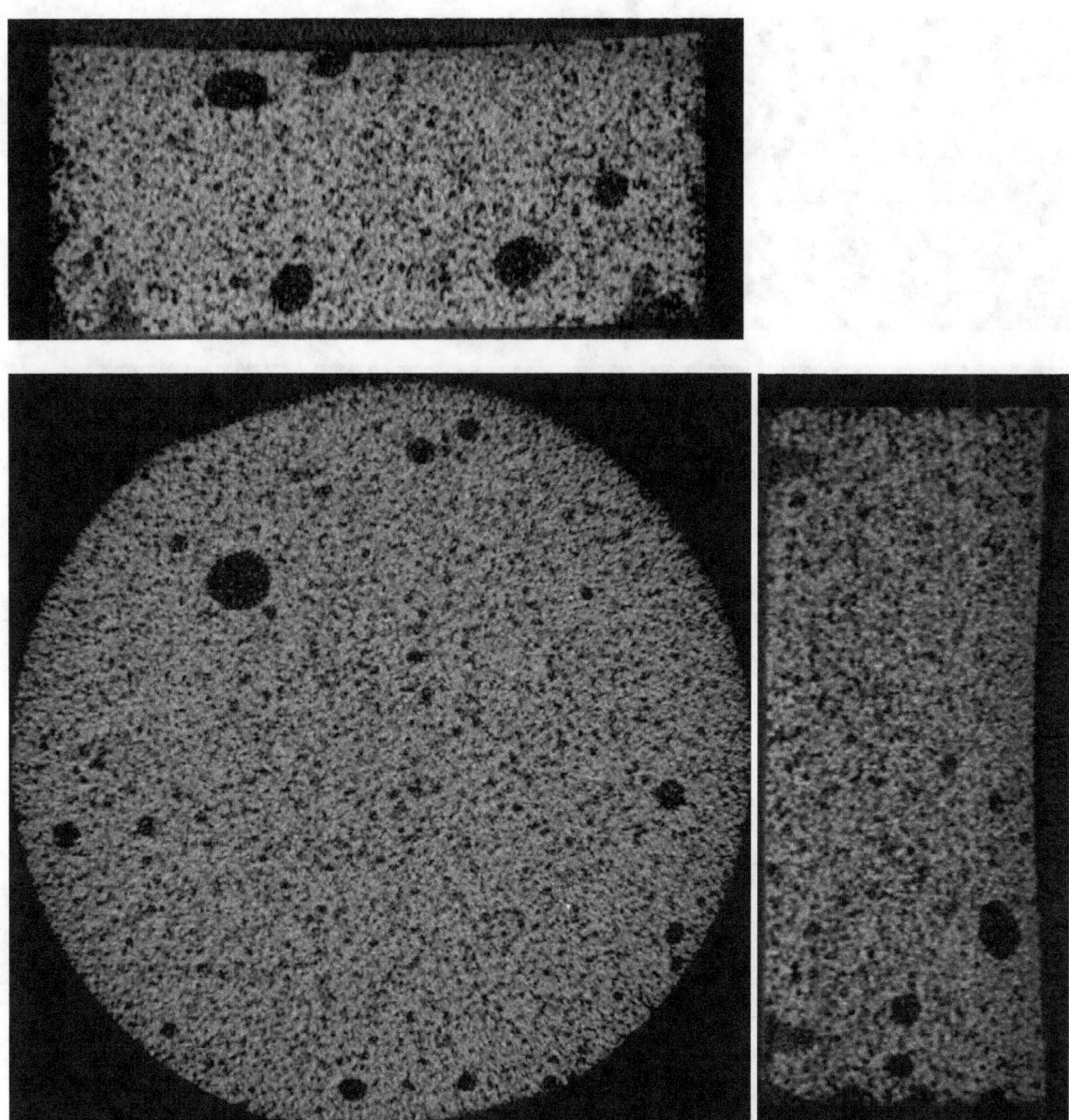

Figure 16. Coronal, transaxial, and sagittal images of mortar #5 after being exposed to drying for 950 min.

Mortar #6 (Accelerator) at 30 min

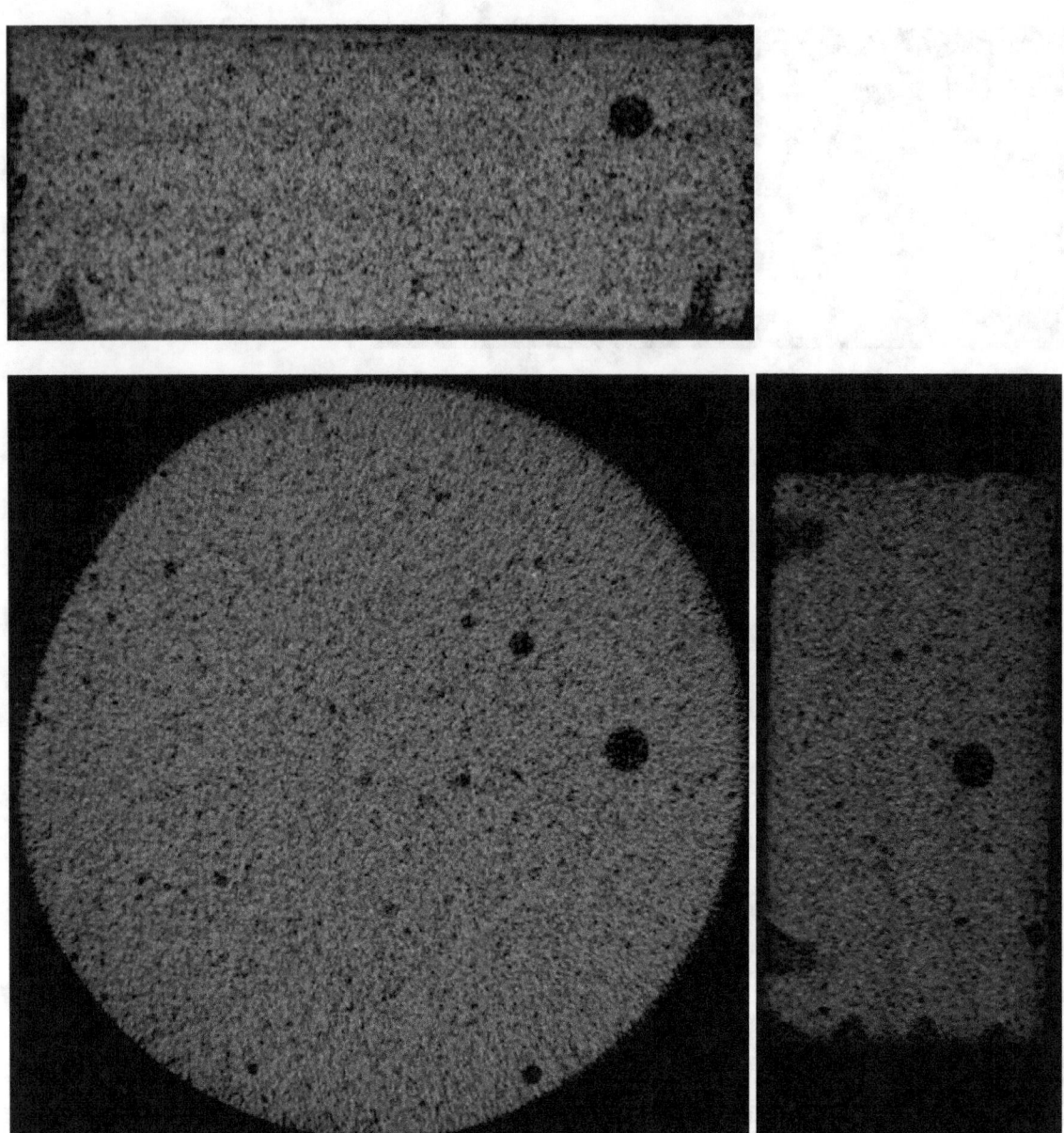

Figure 17. Coronal, transaxial, and sagittal images of mortar #6 after being exposed to drying for 30 min.

Mortar #6 (Accelerator) at 270 min

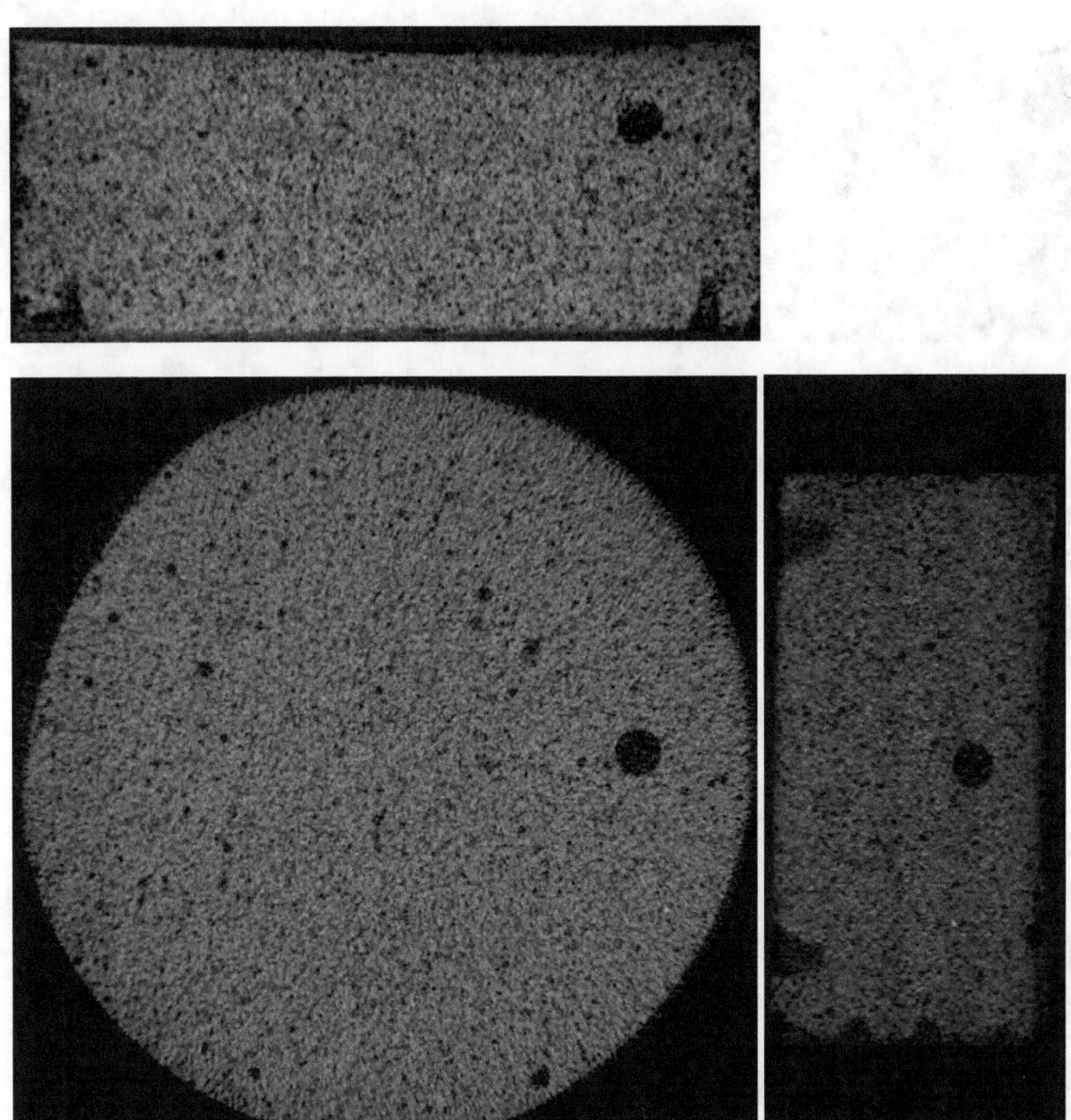

Figure 18. Coronal, transaxial, and sagittal images of mortar #6 after being exposed to drying for 270 min.

Mortar #7 (Polyvinyl alcohol) at 30 min

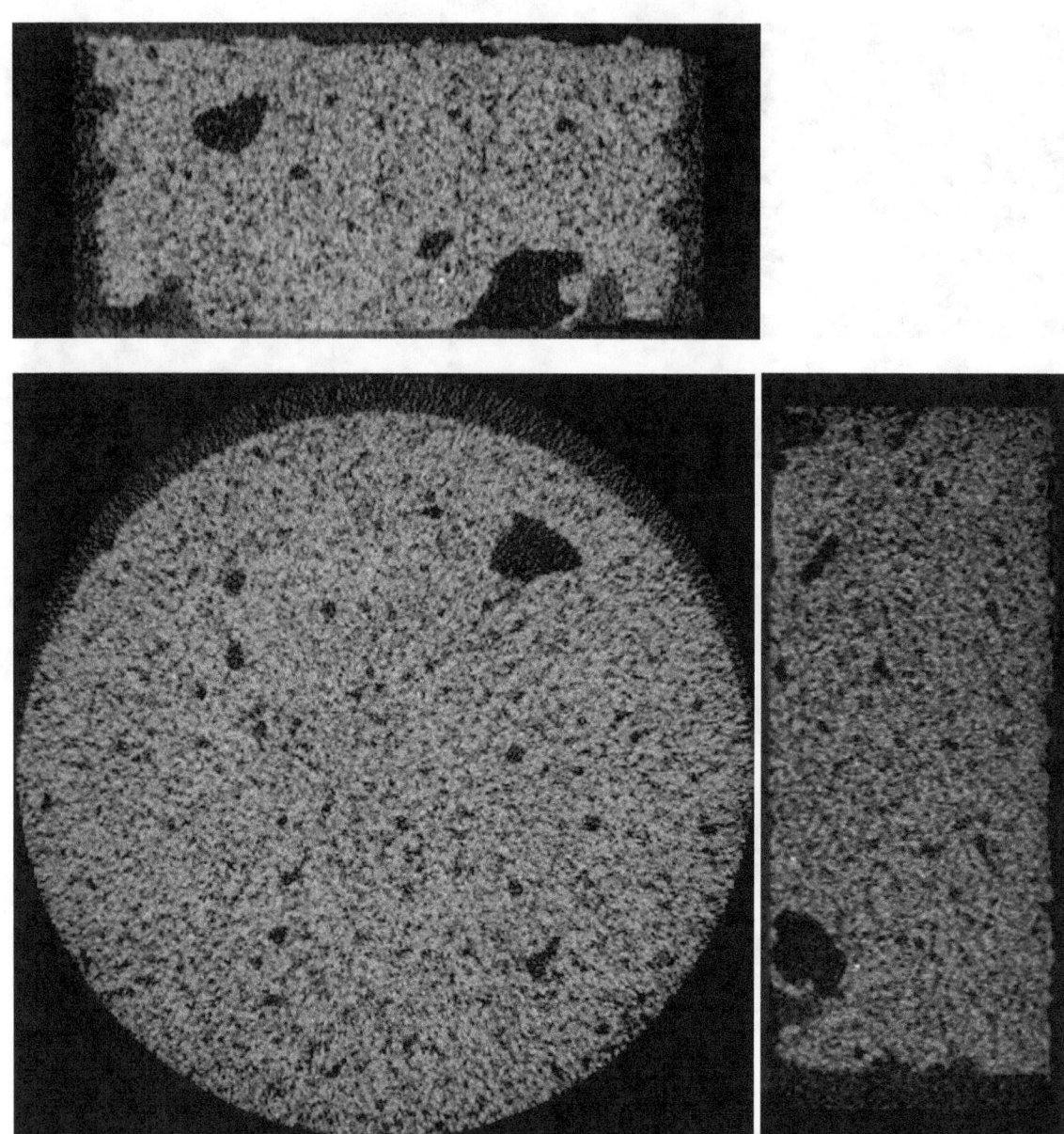

Figure 19. Coronal, transaxial, and sagittal images of mortar #7 after being exposed to drying for 30 min.

Mortar #7 (Polyvinyl alcohol) at 1040 min

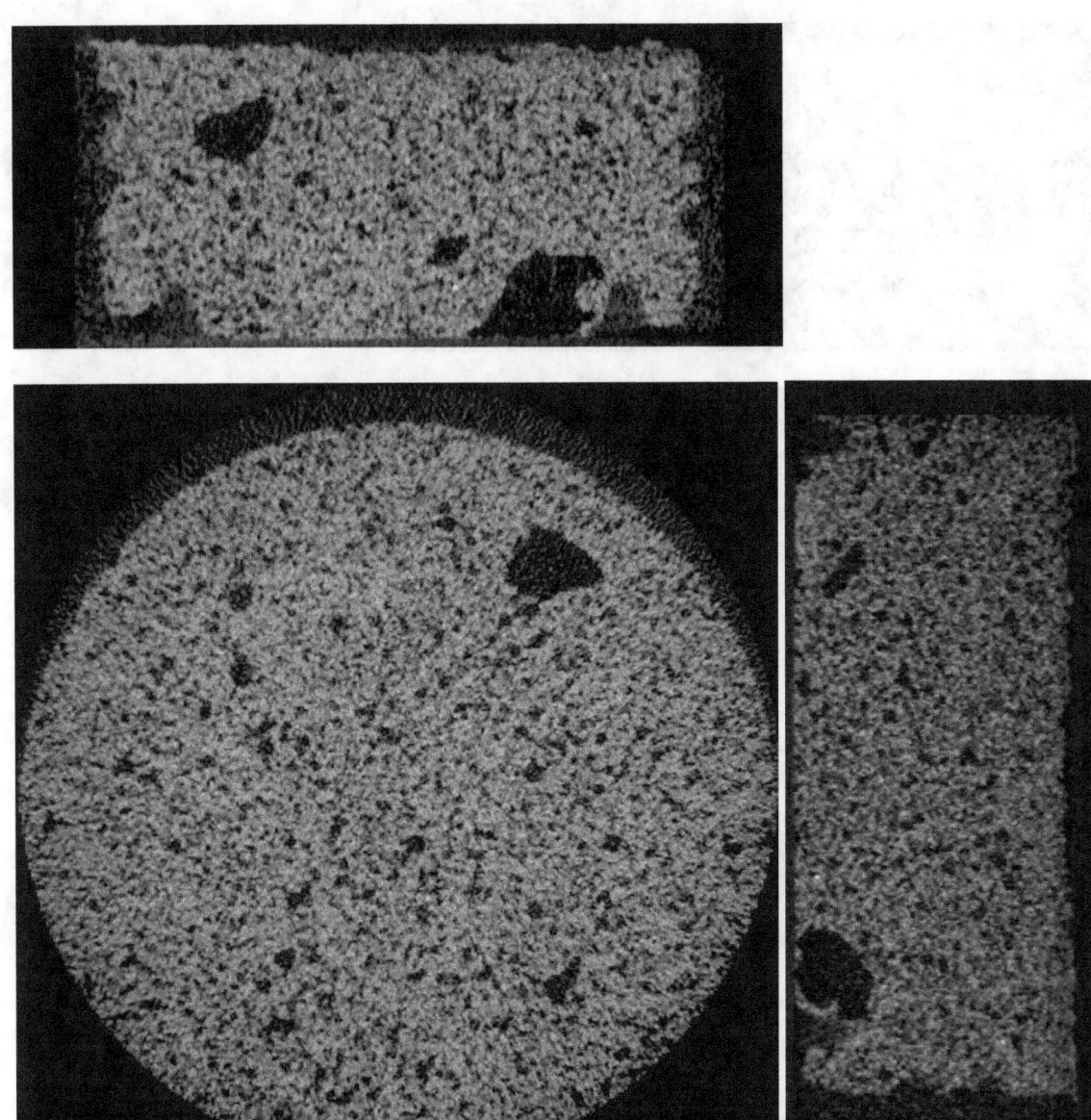

Figure 20. Coronal, transaxial, and sagittal images of mortar #7 after being exposed to drying for 1040 min.

Mortar #8 (Accelerator/Imbentin) at 30 min

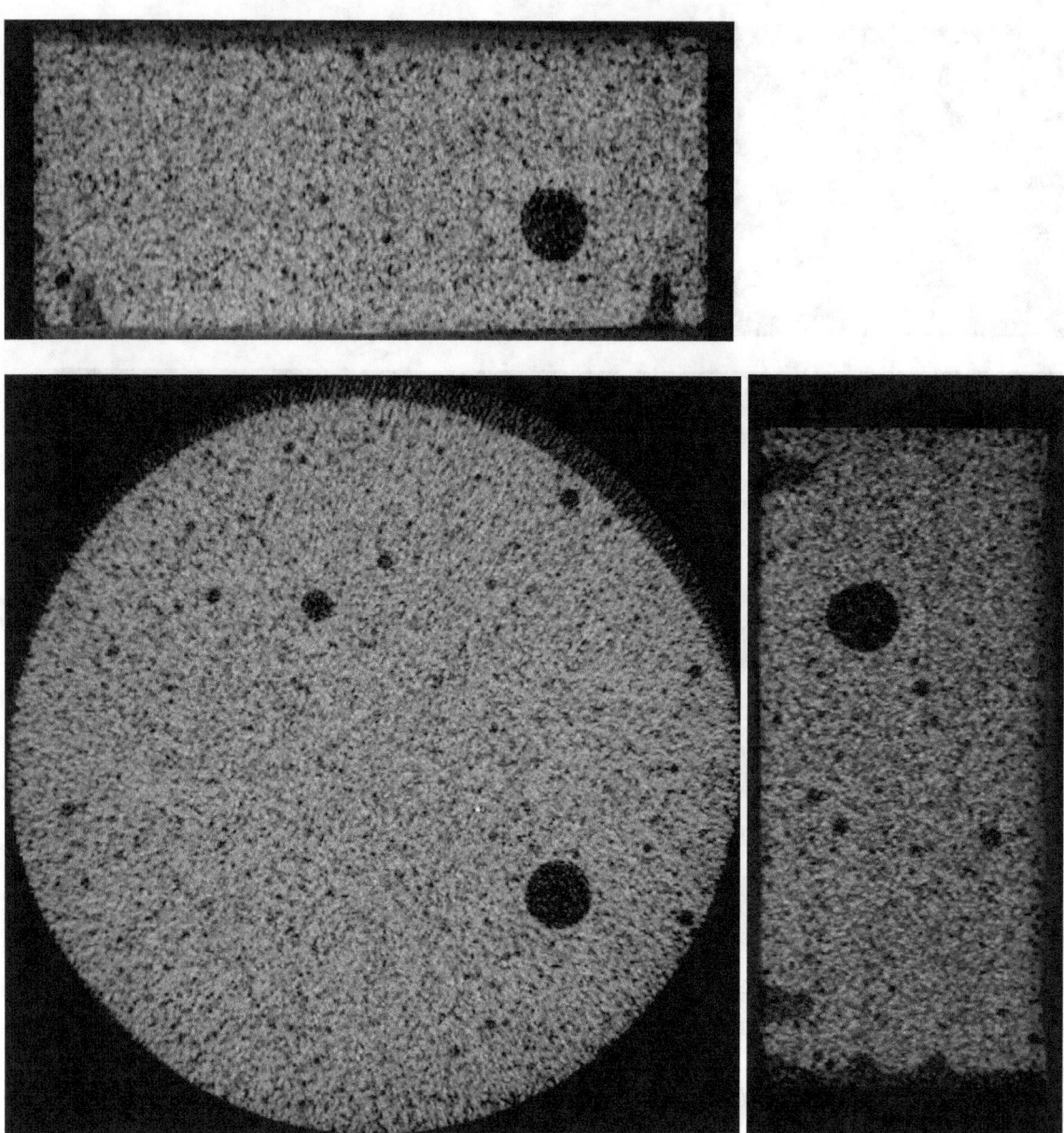

Figure 21. Coronal, transaxial, and sagittal images of mortar #8 after being exposed to drying for 30 min.

Mortar #8 (Accelerator/Imbentin) at 260 min

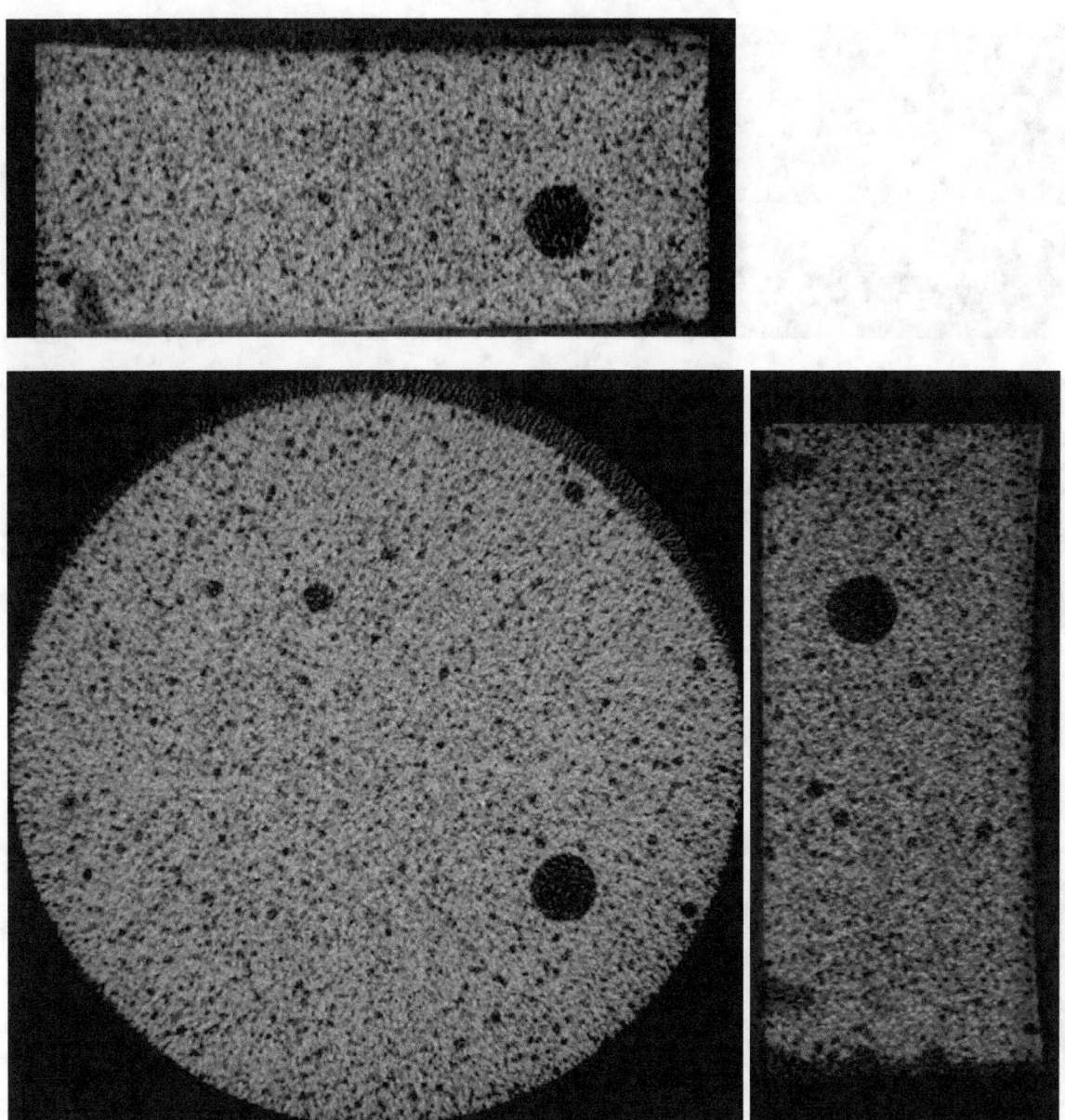

Figure 22. Coronal, transaxial, and sagittal images of mortar #8 after being exposed to drying for 260 min.

Mortar #9 (Starch Ether) at 30 min

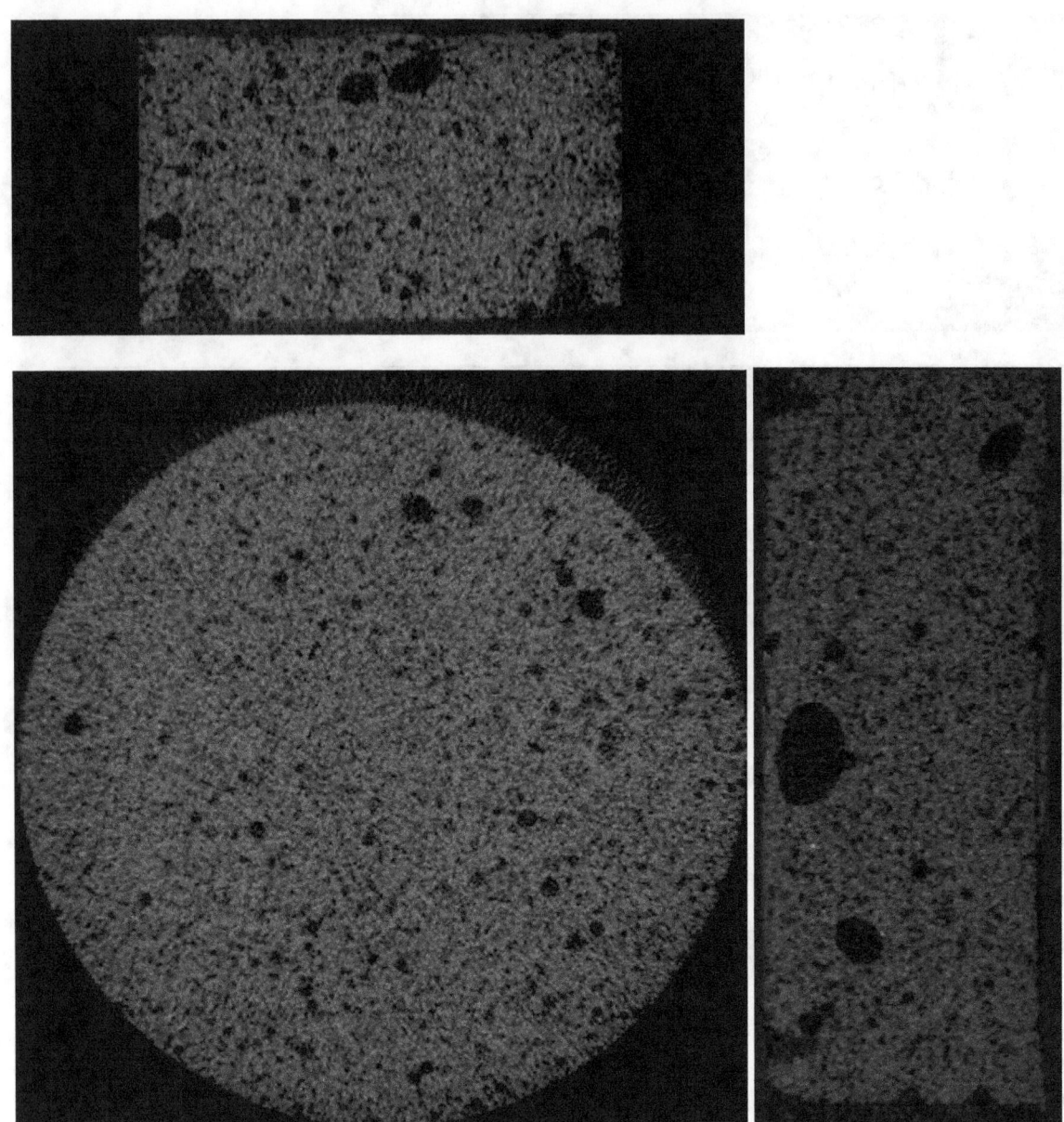

Figure 23. Coronal, transaxial, and sagittal images of mortar #9 after being exposed to drying for 30 min.

Mortar #9 (Starch Ether) at 1370 min

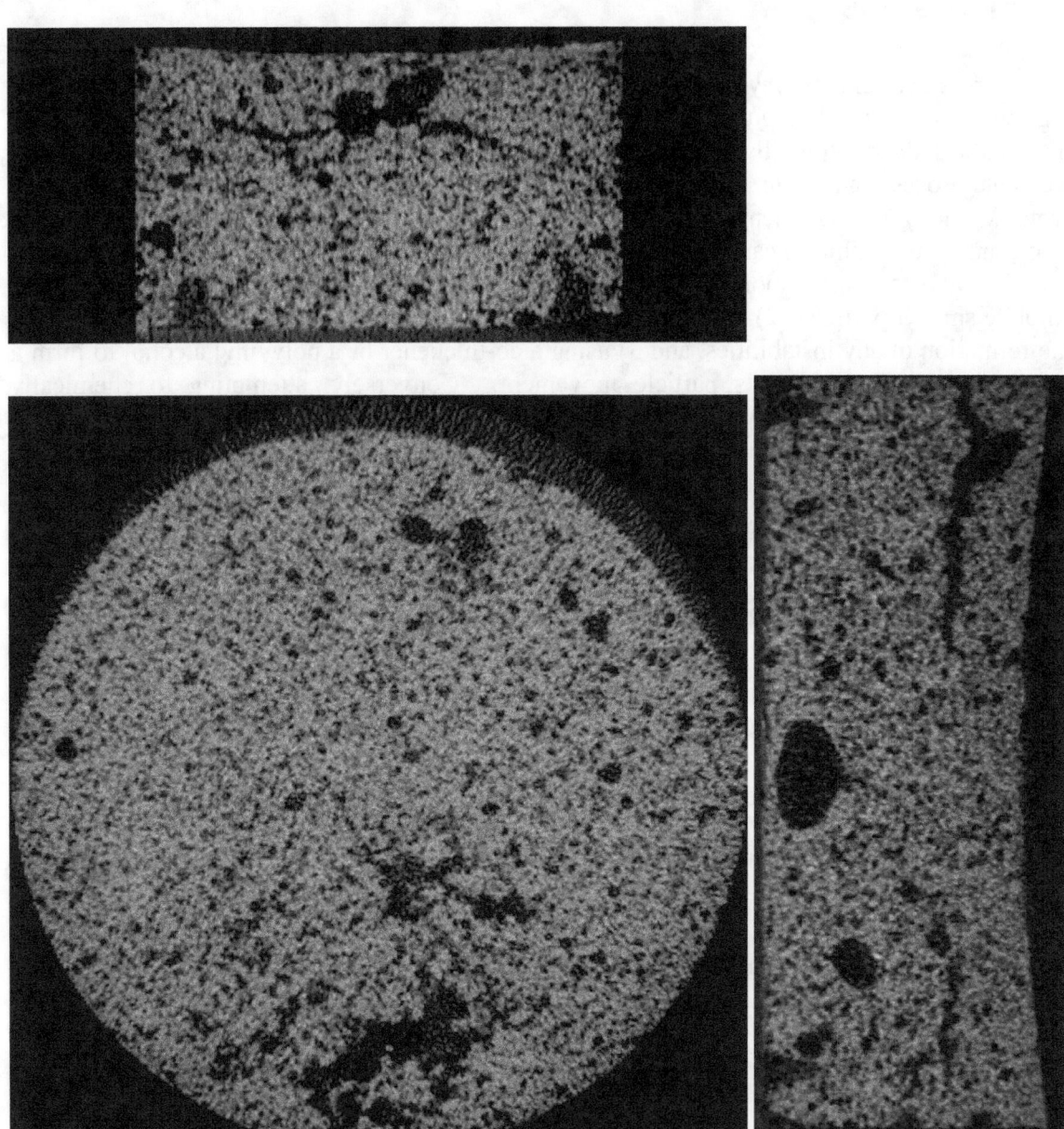

Figure 24. Coronal, transaxial, and sagittal images of mortar #9 after being exposed to drying for 1370 min.

5. Conclusions

X-ray microtomography measurements have verified the occurrence of air void instability and growth during the drying/curing of tile adhesive mortars. Time progression images have clearly indicated that initially stable air voids can destabilize during the drying/particle movement process and can grow to several times their initial volumes. The X-ray microtomography measurements were used (in conjunction with isothermal calorimetry, air content, and open time measurements) to evaluate various mitigation strategies. Three potentially successful mitigation strategies were identified: 1) using a coarser cement to remove the mobile smaller particles, 2) incorporating an accelerator to achieve setting of the mortar prior to the formation of any instabilities, and 3) using a co-thickener or a polyvinyl alcohol to form a polymer network that prevents particle movement. Conversely, attempting to chemically stabilize the air void system by incorporating an air entrainer into the control mixture was not successful in avoiding the formation and growth of unstable voids. X-ray microtomography has demonstrated itself to be a valuable tool for the in situ observation of dynamic processes within hardening cement-based material microstructures.

6. Acknowledgements

C.J. Haecker would like to thank the staff of the Inorganic Materials Group at the National Institute of Standards and Technology (NIST) for hosting a visit during which these experiments were completed. A thorough review of the report by Mr. Duane Emmett and his colleagues at Bostik, Inc. is also greatly appreciated.

References

1) Bentz, D.P., Haecker, C.-J., Peltz, M.A., and Snyder, K.A., "X-ray Absorption Studies of Drying of Cementitious Tile Adhesive Mortars," *Cement and Concrete Composites*, **30** (5), 361-373, 2008.
2) Jenni, A., Holzer, L., Zurbriggen, R., and Herwegh, M., "Influence of Polymers on Microstructure and Adhesive Strength of Cementitious Tile Adhesive Mortars," *Cement and Concrete Research*, **35** (1), 35-50, 2005.
3) Oberste-Padtberg, R., and Sieksmeier, J., "Factors Influencing the Open Time of Building Mortars," Drymix Mortar Yearbook, 44-49, 2007.
4) Bentz, D.P., "Replacement of "Coarse" Cement Particles by Inert Fillers in Low w/c Ratio Concretes II: Experimental Validation," *Cement and Concrete Research*, **35** (1), 185-188, 2005.
5) EN 1347, "Adhesives for tiles - Determination of wetting capability," Comité Européen de Normalisation Technical Committee 67, 1999.

www.ingramcontent.com/pod-product-compliance
Lightning Source LLC
Chambersburg PA
CBHW081804170526
45167CB00008B/3322